Student Edition

Eureka Math
Grade 2
Modules 1, 2, & 3

Special thanks go to the Gordon A. Cain Center and to the Department of Mathematics at Louisiana State University for their support in the development of *Eureka Math*.

For a free *Eureka Math* Teacher
Resource Pack, Parent Tip
Sheets, and more please
visit www.Eureka.tools

ISBN 978-1-63255-293-8

Name _____   Date _____

1. Add or subtract. Complete the number bond for each set.

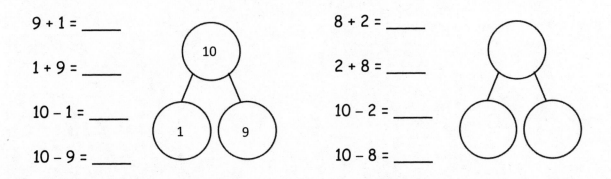

9 + 1 = _____

1 + 9 = _____

10 – 1 = _____

10 – 9 = _____

8 + 2 = _____

2 + 8 = _____

10 – 2 = _____

10 – 8 = _____

2. Solve. Draw a number bond for each set.

6 + 4 = _____

4 + 6 = _____

10 – 4 = _____

10 – 6 = _____

3 + 7 = _____

7 + 3 = _____

10 – 7 = _____

10 – 3 = _____

3. Solve.

10 = 7 + _____

10 = 3 + _____

10 = 5 + _____

10 = 2 + _____

10 = _____ + 8

10 = _____ + 4

10 = _____ + 6

10 = _____ + 1

EUREKA
MATH™

©2015 Great Minds. eureka-math.org
G2-M1-SE-B1-1.3.1-1.2016

This page intentionally left blank

Name _____   Date _____

1. Add or subtract. Draw a number bond for (b).

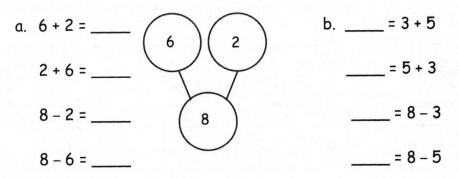

a. 6 + 2 = _____   b. _____ = 3 + 5

    2 + 6 = _____   _____ = 5 + 3

    8 – 2 = _____   _____ = 8 – 3

    8 – 6 = _____   _____ = 8 – 5

2. Solve.

    20 + 4 = _____   _____ = 20 + 9

    40 + 3 = _____   _____ = 40 + 8

    70 + 2 = _____   _____ = 50 + 6

    80 + 5 = _____   _____ = 90 + 7

3. Solve.

    14 = 10 + _____   19 = _____ + 9

    23 = 20 + _____   29 = _____ + 9

    71 = 70 + _____   78 = _____ + 8

    82 = 80 + _____   87 = _____ + 7

Name _____ Date _____

Number Bond Dash

Do as many as you can in 90 seconds. Write the number of bonds you finished here:

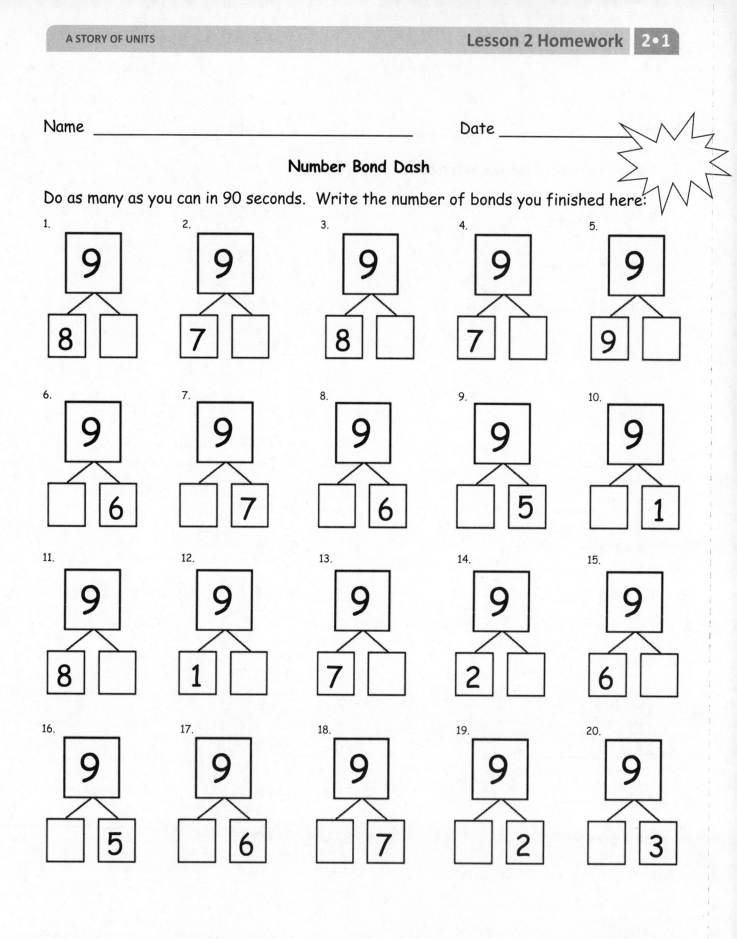

1. 9 / 8, ☐

2. 9 / 7, ☐

3. 9 / 8, ☐

4. 9 / 7, ☐

5. 9 / 9, ☐

6. 9 / ☐, 6

7. 9 / ☐, 7

8. 9 / ☐, 6

9. 9 / ☐, 5

10. 9 / ☐, 1

11. 9 / 8, ☐

12. 9 / 1, ☐

13. 9 / 7, ☐

14. 9 / 2, ☐

15. 9 / 6, ☐

16. 9 / ☐, 5

17. 9 / ☐, 6

18. 9 / ☐, 7

19. 9 / ☐, 2

20. 9 / ☐, 3

Lesson 2: Practice making the next ten and adding to a multiple of ten.

EUREKA MATH

Name _____   Date _____

1. Solve.

    a. 30 + 6 = _____   b. 50 – 30 = _____

       30 + 60 = _____   51 – 30 = _____

       35 + 40 = _____   57 – 4 = _____

       35 + 4 = _____   57 – 40 = _____

2. Solve.

a. 24 + 5 = _____	b. 24 + 50 = _____
c. 78 – 3 = _____	d. 78 – 30 = _____

©2015 Great Minds. eureka-math.org
G2-M1-SE-B1-1.3.1-1.2016

3. Solve.

a. $38 + 10 = \underline{\hspace{1.5cm}}$ $18 + 30 = \underline{\hspace{1.5cm}}$	b. $35 - 10 = \underline{\hspace{1.5cm}}$ $35 - 20 = \underline{\hspace{1.5cm}}$
c. $56 + 40 = \underline{\hspace{1.5cm}}$ $46 + 50 = \underline{\hspace{1.5cm}}$	d. $75 - 40 = \underline{\hspace{1.5cm}}$ $75 - 30 = \underline{\hspace{1.5cm}}$

4. Compare $57 - 2$ to $57 - 20$. How are they different? Use words, drawings, or numbers to explain.

Extension!

5. Andy had $28. He spent $5 on a book.

Lisa had $20 and got $3 more.

Lisa says she has more money.

Prove her right or wrong using pictures, numbers, or words.

Name _____ Date _____

1. Solve.

    a. 20 + 7 = _____   b. 80 – 20 = _____

        20 + 70 = _____   85 – 2 = _____

        62 + 3 = _____   85 – 20 = _____

        62 + 30 = _____   86 – 20 = _____

    c. 30 + 40 = _____   d. 70 – 30 = _____

        31 + 40 = _____   75 – 30 = _____

        35 + 4 = _____   78 – 3 = _____

        45 + 30 = _____   75 – 40 = _____

©2015 Great Minds. eureka-math.org
G2-M1-SE-B1-1.3.1-1.2016

2. Solve.

a. 42 + 7 = _____	b. 24 + 70 = _____
c. 49 – 2 = _____	d. 98 – 20 = _____

3. Solve.

a. 16 + 3 = _____ 13 + 6 = _____	b. 37 – 3 = _____ 37 – 4 = _____
c. 26 + 70 = _____ 76 + 20 = _____	d. 97 – 50 = _____ 97 – 40 = _____

©2015 Great Minds. eureka-math.org
G2-M1-SE-B1-1.3.1-1.2016

Name _____   Date _____

Solve.

1. 9 + 3 = _____	2. 9 + 5 = _____
3. 8 + 4 = _____	4. 8 + 7 = _____
5. 7 + 5 = _____	6. 7 + 6 = _____
7. 8 + 8 = _____	8. 9 + 8 = _____

Solve.

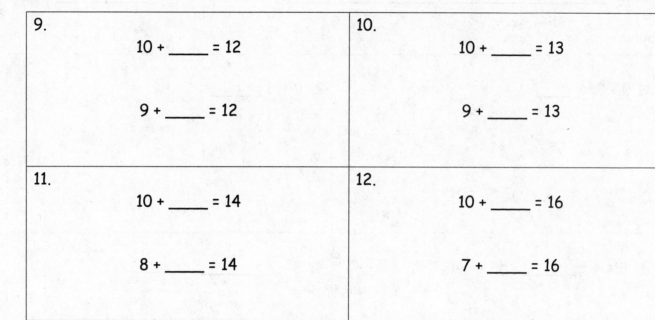

9. 10 + _____ = 12 9 + _____ = 12	10. 10 + _____ = 13 9 + _____ = 13
11. 10 + _____ = 14 8 + _____ = 14	12. 10 + _____ = 16 7 + _____ = 16

13. Lisa has 2 blue beads and 9 purple beads. How many beads does Lisa have in all?

Lisa has _____ beads in all.

14. Ben had 8 pencils and bought 5 more. How many pencils does Ben have altogether?

EUREKA MATH™

©2015 Great Minds. eureka-math.org
G2-M1-SE-B1-1.3.1-1.2016

Name _____   Date _____

Solve.

1. 8 + 4 = _____
     / \\
  2 2
 	
         8 + 2 = 10
      10 + 2 = 12
	2. 9 + 7 = _____

3. 9 + 3 = _____	4. 8 + 6 = _____
5. 7 + 6 = _____	6. 7 + 8 = _____
7. 8 + 8 = _____	8. 8 + 9 = _____

9. Solve and match.

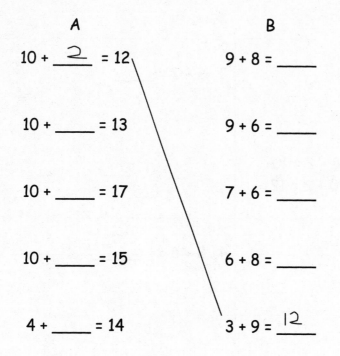

A

$10 + \underline{2} = 12$

$10 + \underline{\hspace{1cm}} = 13$

$10 + \underline{\hspace{1cm}} = 17$

$10 + \underline{\hspace{1cm}} = 15$

$4 + \underline{\hspace{1cm}} = 14$

B

$9 + 8 = \underline{\hspace{1cm}}$

$9 + 6 = \underline{\hspace{1cm}}$

$7 + 6 = \underline{\hspace{1cm}}$

$6 + 8 = \underline{\hspace{1cm}}$

$3 + 9 = \underline{12}$

10. Ronnie uses 5 brown bricks and 8 red bricks to build a fort.
    How many bricks does Ronnie use in all?

Ronnie uses _____ bricks.

Name _____   Date _____

1. Solve.

a. 9 + 3 = _____ $\wedge$ 1 2	b. 19 + 3 = _____
c. 18 + 4 = _____	d. 38 + 7 = _____
e. 37 + 5 = _____	f. 57 + 6 = _____
g. 6 + 68 = _____	h. 8 + 78 = _____

2. Maria solved 67 + 5 as shown. Show Maria a faster way to solve 67 + 5.

$$67 + 5 = 72$$

3. Use the RDW process to solve.

   Jessa collected 78 shells on the beach.
   Susan collected 6 more shells than Jessa.
   How many shells did Susan collect?

Name _____ Date _____

1. Solve.

a. 9 + 3 = _____  1 2	b. 29 + 5 = _____
c. 49 + 7 = _____	d. 59 + 6 = _____
e. 18 + 4 = _____	f. 48 + 6 = _____
g. 58 + 6 = _____	h. 78 + 8 = _____

©2015 Great Minds. eureka-math.org
G2-M1-SE-B1-1.3.1-1.2016

2. Solve.

a. 67 + 5 = _____	b. 87 + 6 = _____
c. 6 + 59 = _____	d. 7 + 78 = _____

3. Use the RDW process to solve.

There were 28 students at recess. A group of 7 students came outside to join them. How many students are there now?

©2015 Great Minds. eureka-math.org
G2-M1-SE-B1-1.3.1-1.2016

Name _____   Date _____

1. Solve.

a. 20 − 9 = _____
     / \
  10 10
           10 − 9 = 1
           10 + 1 = 11
	b. 30 − 9 = _____

c. 20 − 8 = _____	d. 30 − 7 = _____
e. 40 − 7 = _____	f. 50 − 6 = _____
g. 80 − 6 = _____	h. 90 − 5 = _____

Lesson 6:   Subtract single-digit numbers from multiples of 10 within 100.

17

©2015 Great Minds. eureka-math.org
G2-M1-SE-B1-1.3.1-1.2016

i. 70 – 4 = _____	j. 60 – 2 = _____

2. Fill in the number bond and solve.

90 – 9 = ___

/\

_____   _____

3. Show how 10 – 6 helps you solve 50 – 6.

4. Carla has 70 paper clips.

   She gives 6 away.

   How many paper clips does Carla have left?

   Carla has _____ paper clips left.

Name _____   Date _____

1. Take out ten.

30 / \ 20 10	40	50
70	60	80

2. Solve.

10 – 1 = _____	10 – 4 = _____	10 – 9 = _____
10 – 7 = _____	10 – 2 = _____	10 – 5 = _____

3. Solve.

a. 20 – 9 = _11_
        / \
     10   10
 	
                  10 – 9 = 1
                  10 + 1 = 11
	b. 30 – 9 = _____

c. 40 – 8 = _____	d. 50 – 8 = _____
e. 60 – 7 = _____	f. 70 – 7 = _____
g. 80 – 6 = _____	h. 90 – 5 = _____

4. Show how 10 – 4 helps you solve 30 – 4.

©2015 Great Minds. eureka-math.org
G2-M1-SE-B1-1.3.1-1.2016

Name _____ Date _____

1. Solve.

a. 11 − 9 = ____ /\ 1 10	**b.** 12 − 9 = ____	**c.** 13 − 9 = ____
d. 11 − 8 = ____	**e.** 12 − 8 = ____	**f.** 13 − 8 = ____
g. 11 − 7 = ____	**h.** 12 − 7 = ____	**i.** 13 − 7 = ____

2. Solve.

a.	b.	c.
14 – 6 = _____	11 – 5 = _____	16 – 7 = _____

Solve.

3. Shane has 12 pencils. He gives some pencils to his friends. Now, he has 7 left. How many pencils did he give away?

_____

4. Victoria gave 6 celery sticks to her mom. She started with 13. How many celery sticks does she have left?

_____

EUREKA MATH

Name _____   Date _____

1. Take out ten.

17 / \\ 7 10	14	18
13	16	19

2. Solve.

10 – 2 = _____	10 – 7 = _____	10 – 6 = _____
10 – 5 = _____	10 – 8 = _____	10 – 9 = _____

3. Solve.

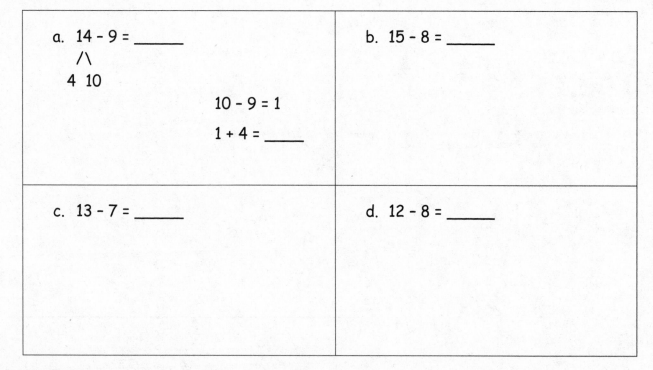

a. 14 – 9 = _____
      /\
    4 10

          10 – 9 = 1

          1 + 4 = _____

b. 15 – 8 = _____

c. 13 – 7 = _____

d. 12 – 8 = _____

Solve.

4. Robert has 16 cups. Some are red. Nine are blue. How many cups are red?

_____ cups are red.

5. Lucy spent $8 on a game. She started with $14. How much money does Lucy have left?

_____

Name _____ Date _____

1. Solve.

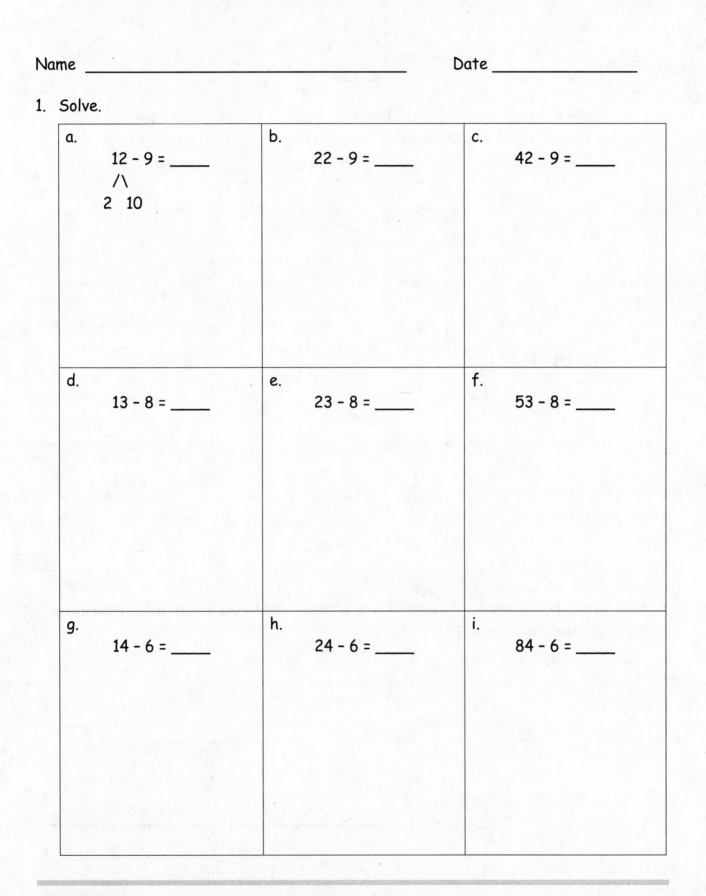

a.	b.	c.
12 - 9 = ____ /\ 2 10	22 - 9 = ____	42 - 9 = ____
d. 13 - 8 = ____	e. 23 - 8 = ____	f. 53 - 8 = ____
g. 14 - 6 = ____	h. 24 - 6 = ____	i. 84 - 6 = ____

2. Solve.

a.	b.	c.
24 – 9 = _____	36 – 7 = _____	53 – 6 = _____
d.	e.	f.
42 – 8 = _____	61 – 5 = _____	85 – 8 = _____

3. Mrs. Watts had 17 tacos. The children ate some. Nine tacos were left. How many tacos did the children eat?

_____

Lesson 8:   Take from 10 within 100.

Name _____   Date _____

1. Take out ten.

26 / \\ 16 10	34	58
85	77	96

2. Solve.

10 – 1 = _____	10 – 5 = _____	10 – 2 = _____
10 – 4 = _____	10 – 7 = _____	10 – 8 = _____

3. Solve.

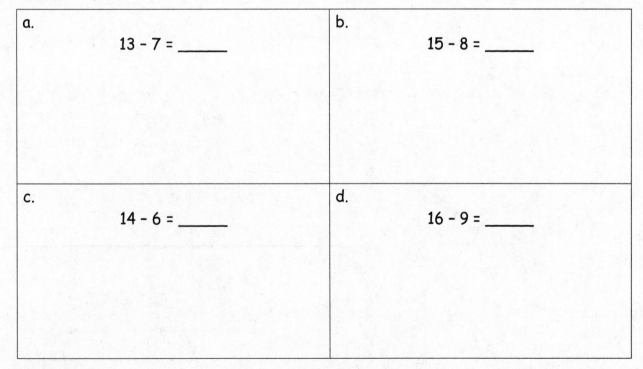

a. 13 – 7 = _____	b. 15 – 8 = _____
c. 14 – 6 = _____	d. 16 – 9 = _____

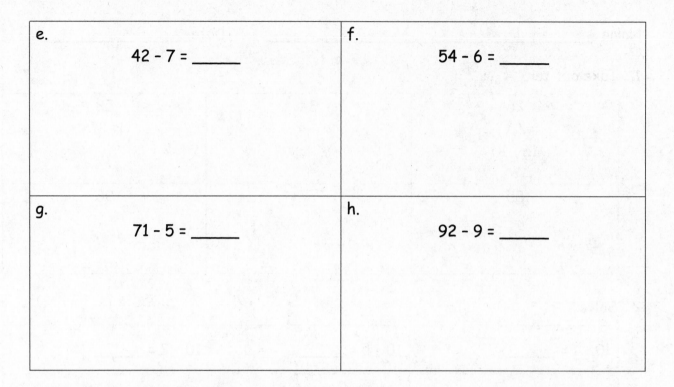

e.

42 – 7 = _____

f.

54 – 6 = _____

g.

71 – 5 = _____

h.

92 – 9 = _____

4. Emma has 16 markers. She gave Jack some. Seven markers are left. How many markers did Emma give Jack?

_____

Lesson 8:   Take from 10 within 100.

Eureka Math
Grade 2
Module 2

Special thanks go to the Gordon A. Cain Center and to the Department of Mathematics at Louisiana State University for their support in the development of *Eureka Math*.

For a free *Eureka Math* Teacher
Resource Pack, Parent Tip
Sheets, and more please
visit www.Eureka.tools

Published by the non-profit Great Minds

Copyright © 2015 Great Minds. No part of this work may be reproduced, sold, or commercialized, in whole or in part, without written permission from Great Minds. Non-commercial use is licensed pursuant to a Creative Commons Attribution-NonCommercial-ShareAlike 4.0 license; for more information, go to http://greatminds.net/maps/math/copyright. "Great Minds" and "Eureka Math" are registered trademarks of Great Minds.

Printed in the U.S.A.
This book may be purchased from the publisher at eureka-math.org
10 9 8

ISBN 978-1-63255-293-8

Name _____ Date _____

Use centimeter cubes to find the length of each object.

1. The picture of the fork and spoon is about _____ centimeter cubes long.

2. The picture of the hammer is about _____ centimeters long.

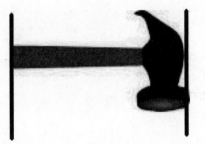

3. The length of the picture of the comb is about _____ centimeters.

Lesson 1:   Connect measurement with physical units by using multiple copies of the same physical unit to measure.

1

©2015 Great Minds. eureka-math.org
G2-M2-SE-B1-1.3.1-1.2016

4. The length of the picture of the shovel is about _____ centimeters.

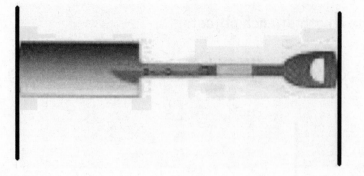

5. The head of a grasshopper is 2 centimeters long. The rest of the grasshopper's body is 7 centimeters long. What is the total length of the grasshopper?

6. The length of a screwdriver is 19 centimeters. The handle is 5 centimeters long.

   a. What is the length of the top of the screwdriver?

   b. How much shorter is the handle than the top of the screwdriver?

Lesson 1:   Connect measurement with physical units by using multiple copies of the same physical unit to measure.

Name _____   Date _____

Count each centimeter cube to find the length of each object.

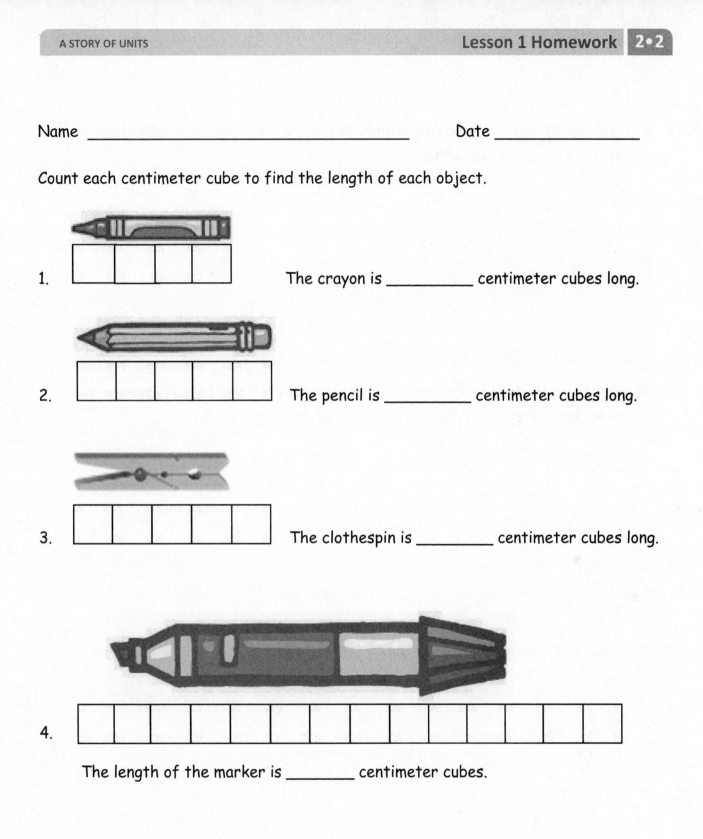

1.   The crayon is _____ centimeter cubes long.

2.   The pencil is _____ centimeter cubes long.

3.   The clothespin is _____ centimeter cubes long.

4.

The length of the marker is _____ centimeter cubes.

Lesson 1:   Connect measurement with physical units by using multiple copies of the same physical unit to measure.   **3**

©2015 Great Minds. eureka-math.org
G2-M2-SE-B1-1.3.1-1.2016

5. Richard has 43 centimeter cubes. Henry has 30 centimeter cubes. What is the length of their cubes altogether?

6. The length of Marisa's loaf of bread is 54 centimeters. She cut off and ate 7 centimeters of bread. What is the length of what she has left?

7. The length of Jimmy's math book is 17 centimeter cubes. His reading book is 12 centimeter cubes longer. What is the length of his reading book?

©2015 Great Minds. eureka-math.org
G2-M2-SE-B1-1.3.1-1.2016

Name _____   Date _____

Find the length of each object using one centimeter cube. Mark the endpoint of each centimeter cube as you measure.

1. The picture of the eraser is about _____ centimeters long.

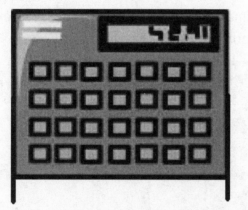

2. The picture of the calculator is about _____ centimeters long.

3. The length of the picture of the envelope is about _____ centimeters.

Lesson 2:   Use iteration with one physical unit to measure.

5

©2015 Great Minds. eureka-math.org
G2-M2-SE-B1-1.3.1-1.2016

4. Jayla measured her puppet's legs to be 23 centimeters long. The stomach is 7 centimeters long, and the neck and head together are 10 centimeters long. What is the total length of the puppet?

5. Elijah begins measuring his math book with his centimeter cube. He marks off where each cube ends. After a few times, he decides this process is taking too long and starts to guess where the cube would end and then mark it.

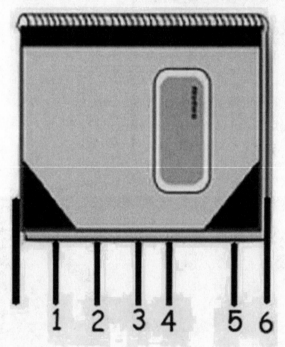

Explain why Elijah's answer will be incorrect.

_____

_____

Name _____   Date _____

Use the centimeter square at the bottom of the next page to measure the length of each object. Mark the endpoint of the square as you measure.

1. The picture of the glue is about _____ centimeters long.

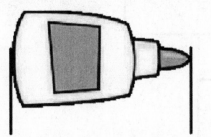

2. The picture of the lollipop is about _____ centimeters long.

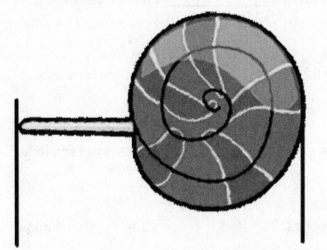

3. The picture of the scissors is about _____ centimeters long.

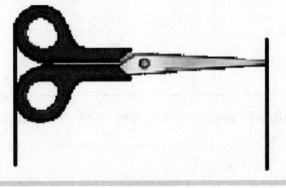

4. Samantha used a centimeter cube and the mark and move forward strategy to measure these ribbons. Use her work to answer the following questions.

Red Ribbon

Blue Ribbon

Yellow Ribbon

a. How long is the red ribbon? _____ centimeters long.

b. How long is the blue ribbon? _____ centimeters long.

c. How long is the yellow ribbon? _____ centimeters long.

d. Which ribbon is the longest?   Red   Blue   Yellow

e. Which ribbon is the shortest?   Red   Blue   Yellow

f. The total length of the ribbons is _____ centimeters.

Cut out the centimeter square below to measure the length of the glue bottle, lollipop, and scissors.

EUREKA MATH™

Name _____   Date _____

Use your centimeter ruler to measure the length of the objects below.

1. The picture of the animal track is about _____ cm long.

2. The picture of the turtle is about _____ cm long.

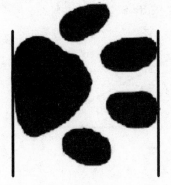

3. The picture of the sandwich is about _____ cm long.

Lesson 3:   Apply concepts to create unit rulers and measure lengths using unit rulers.

©2015 Great Minds. eureka-math.org
G2-M2-SE-B1-1.3.1-1.2016

9

4. Measure and label the length of each side of the triangle using your ruler.

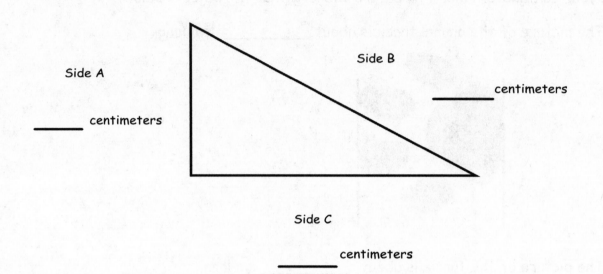

Side A

_____ centimeters

Side B

_____ centimeters

Side C

_____ centimeters

a. Which side is the shortest?   Side A   Side B   Side C

b. What is the length of Sides A and B together? _____ centimeters

c. How much shorter is Side C than Side B? _____ centimeters

Apply concepts to create unit rulers and measure lengths using unit rulers.

Name _____   Date _____

1. Measure five things in the classroom with a centimeter ruler. List the five things and their length in centimeters.

Object Name	Length in Centimeters
a.	
b.	
c.	
d.	
e.	

2. Measure four things in the classroom with a meter stick or meter tape. List the four things and their length in meters.

Object Name	Length in Meters
a.	
b.	
c.	
d.	

©2015 Great Minds. eureka-math.org
G2-M2-SE-B1-1.3.1-1.2016

3. List five things in your house that you would measure with a meter stick or meter tape.

   a. _____

   b. _____

   c. _____

   d. _____

   e. _____

   Why would you want to measure those five items with a meter stick or meter tape instead of a centimeter ruler?

   _____

   _____

   _____

4. The distance from the cafeteria to the gym is 14 meters. The distance from the cafeteria to the playground is double that distance. How many times would you need to use a meter stick to measure the distance from the cafeteria to the playground?

Name _____   Date _____

1. Circle cm (centimeter) or m (meter) to show which unit you would use to measure the length of each object.

   a. Length of a marker   cm or m

   b. Length of a school bus   cm or m

   c. Length of a laptop computer   cm or m

   d. Length of a highlighter marker   cm or m

   e. Length of a football field   cm or m

   f. Length of a parking lot   cm or m

   g. Length of a cell phone   cm or m

   h. Length of a lamp   cm or m

   i. Length of a supermarket   cm or m

   j. Length of a playground   cm or m

2. Fill in the blanks with **cm** or **m**.

   a. The length of a swimming pool is 25 _____.

   b. The height of a house is 8 _____.

   c. Karen is 6 _____ shorter than her sister.

   d. Eric ran 65 _____ down the street.

   e. The length of a pencil box is 3 _____ longer than a pencil.

Lesson 4:   Measure various objects using centimeter rulers and meter sticks.

©2015 Great Minds. eureka-math.org
G2-M2-SE-B1-1.3.1-1.2016

3. Use the centimeter ruler to find the length (from one mark to the next) of each object.

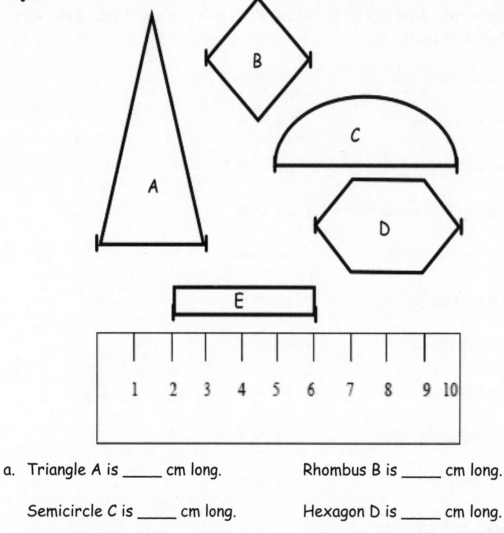

a. Triangle A is _____ cm long.   Rhombus B is _____ cm long.

Semicircle C is _____ cm long.   Hexagon D is _____ cm long.

Rectangle E is _____ cm long.

b. Explain how the strategy to find the length of each shape above is different from how you would find the length if you used a centimeter cube.

_____

_____

_____

Name _____ Date _____

First, estimate the length of each line in centimeters using mental benchmarks.
Then, measure each line with a centimeter ruler to find the actual length.

1. _____

   a. Estimate: _____ cm   b. Actual length: _____ cm

2. _____

   a. Estimate: _____ cm   b. Actual length: _____ cm

3. _____

   a. Estimate: _____ cm   b. Actual length: _____ cm

4. _____

   a. Estimate: _____ cm   b. Actual length: _____ cm

5. _____

   a. Estimate: _____ cm   b. Actual length: _____ cm

EUREKA
MATH™

Lesson 5:   Develop estimation strategies by applying prior knowledge of length
and using mental benchmarks.

17

©2015 Great Minds. eureka-math.org
G2-M2-SE-B1-1.3.1-1.2016

6. Circle the correct unit of measurement for each length estimate.

   a. The height of a door is about 2 (centimeters/meters) tall.

      What benchmark did you use to estimate? _____

   b. The length of a pen is about 10 (centimeters/meters) long.

      What benchmark did you use to estimate? _____

   c. The length of a car is about 4 (centimeters/meters) long.

      What benchmark did you use to estimate? _____

   d. The length of a bed is about 2 (centimeters/meters) long.

      What benchmark did you use to estimate? _____

   e. The length of a dinner plate is about 20 (centimeters/meters) long.

      What benchmark did you use to estimate? _____

7. Use an unsharpened pencil to estimate the length of 3 things in your desk.

   a. _____ is about _____ cm long.

   b. _____ is about _____ cm long.

   c. _____ is about _____ cm long.

Lesson 5: Develop estimation strategies by applying prior knowledge of length and using mental benchmarks.

EUREKA MATH

Name _____   Date _____

1. Name five things in your home that you would measure in meters.
    Estimate their length.

    *Remember, the length from a doorknob to the floor is about 1 meter.

Item	Estimated Length
a.	
b.	
c.	
d.	
e.	

2. Choose the best length estimate for each object.

    a. Whiteboard   3 m   or   45 cm

    b. Banana   14 cm   or   30 cm

    c. DVD   25 cm   or   17 cm

    d. Pen   16 cm   or   1 m

    e. Swimming pool   50 m   or   150 cm

EUREKA
MATH™

Lesson 5:   Develop estimation strategies by applying prior knowledge of length
              and using mental benchmarks.

19

©2015 Great Minds. eureka-math.org
G2-M2-SE-B1-1.3.1-1.2016

3. The width of your pinky finger is about 1 cm.

   Measure the length of the lines using your pinky finger. Write your estimate.

   a. Line A _____

      Line A is about _____ cm long.

   b. Line B _____

      Line B is about _____ cm long.

   c. Line C

      _____

      Line C is about _____ cm long.

   d. Line D _____

      Line D is about _____ cm long.

   e. Line E _____

      Line E is about _____ cm long.

Lesson 5:   Develop estimation strategies by applying prior knowledge of length
                    and using mental benchmarks.

Name _____   Date _____

Measure each set of lines in centimeters, and write the length on the line. Complete the comparison sentences.

1. Line A ——————————————————————————————————————

    Line B ———————————————————

        a. Line A   Line B

            _____ cm   _____ cm

        b. Line A is about _____ cm longer than Line B.

2. Line C   ————————————————————————

    Line D   ————————————————————————————

        a. Line C   Line D

            _____ cm   _____ cm

        b. Line C is about _____ cm shorter than Line D.

©2015 Great Minds. eureka-math.org
G2-M2-SE-B1-1.3.1-1.2016

3. Line E ———————————

    Line F   ———————————————————

    Line G   ———————————————————

    a. Line E   Line F   Line G

        _____ cm   _____ cm   _____ cm

    b. Lines E, F, and G are about_____ cm combined.

    c. Line E is about _____ cm shorter than Line F.

    d. Line G is about_____ cm longer than Line F.

    e. Line F doubled is about _____ cm longer than Line G.

4. Daniel measured the heights of some young trees in the orchard. He wants to know
    how many more centimeters are needed to have a height of 1 meter. Fill in the blanks.

    a. 90 cm + _____ cm = 1 m

    b. 80 cm + _____ cm = 1 m

    c. 85 cm + _____ cm = 1 m

    d. 81 cm + _____ cm = 1 m

©2015 Great Minds. eureka-math.org
G2-M2-SE-B1-1.3.1-1.2016

5. Carol's ribbon is 76 centimeters long. Alice's ribbon is 1 meter long. How much longer is Alice's ribbon than Carol's?

6. The cricket hopped a distance of 52 centimeters. The grasshopper hopped 9 centimeters farther than the cricket. How far did the grasshopper jump?

7. The pencil box is 24 centimeters in length and 12 centimeters wide. How many more centimeters is the length than the width? _____ more cm

   Draw the rectangle and label the sides.

   What is the total length of all four sides? _____ cm

This page intentionally left blank

Name _____   Date _____

Measure each set of lines in centimeters, and write the length on the line. Complete the comparison sentences.

1. Line A _____

    Line B   _____

    a. Line A is about _____ cm longer than line B.

    b. Line A and B are about _____ cm combined.

2. Line X _____

    Line Y   _____

    Line Z   _____

    a.   Line X   Line Y   Line Z

           _____ cm   _____ cm   _____ cm

    b. Lines X, Y, and Z are about_____ cm combined.

    c. Line Z is about _____ cm shorter than Line X.

    d. Line X is about _____ cm shorter than Line Y.

    e. Line Y is about _____ cm longer than Line Z.

    f. Line X doubled is about _____ cm longer than line Y.

EUREKA
MATH™

3. Line J is 60 cm long. Line K is 85 cm long. Line L is 1 m long.

    a. Line J is _____ cm shorter than line K.

    b. Line L is _____ cm longer than line K.

    c. Line J doubled is _____ cm more than line L.

    d. Lines J, K, and L combined are _____ cm.

4. Katie measured the seat height of four different chairs in her house. Here are her results:

    Loveseat height: 51 cm   Highchair height: 97 cm
    Dining room chair height: 55 cm   Counter stool height: 65 cm

    a. How much shorter is the dining room chair than the counter stool? _____ cm

    b. How much taller is a meter stick than the counter stool? _____ cm

    c. How much taller is a meter stick than the loveseat? _____ cm

5. Max ran 15 meters this morning. This afternoon, he ran 48 meters.

    a. How many more meters did he run in the afternoon?

    b. How many meters did Max run in all?

EUREKA
MATH™

Name _____ Date _____

Measure each set of lines with one small paper clip, using mark and move forward. Measure each set of lines in centimeters using a ruler.

1. Line A ————————————————————

   Line B   ————————————————

   a. Line A

      _____ paper clips   _____ cm

   b. Line B

      _____ paper clips   _____ cm

   c. Line B is about _____ paper clips shorter than Line A.

   d. Line A is about _____ cm longer than Line B.

2.   ———————————————————————— Line L

              ———————— Line M

   a. Line L

      _____ paper clips   _____ cm

   b. Line M

      _____ paper clips   _____ cm

   c. Line L is about _____ paper clips longer than Line M.

   d. Line M doubled is about _____ cm shorter than Line L.

Lesson 7:   Measure and compare lengths using standard metric length units and
            non-standard length units; relate measurement to unit size.

29

©2015 Great Minds. eureka-math.org
G2-M2-SE-B1-1.3.1-1.2016

3. Draw a line that is 6 cm long and another line below it that is 15 cm long. Label the 6 cm line C and the 15 cm line D.

   a. Line C   Line D

        _____ paper clips   _____ paper clips

   b. Line D is about _____ cm longer than Line C.

   c. Line C is about _____ paper clips shorter than Line D.

   d. Lines C and D together are about _____ paper clips long.

   e. Lines C and D together are about _____ centimeters long.

4. Christina measured Line F with quarters and Line G with pennies.

Line F is about 6 quarters long. Line G is about 8 pennies long. Christina said Line G is longer because 8 is a bigger number than 6.

Explain why Christina is incorrect.

_____

_____

Lesson 7:   Measure and compare lengths using standard metric length units and non-standard length units; relate measurement to unit size.

©2015 Great Minds. eureka-math.org
G2-M2-SE-B1-1.3.1-1.2016

Name _____ Date _____

Use a centimeter ruler and paper clips to measure and compare lengths.

1. ————————————————————— Line Z

  a. Line Z

      _____ paper clips   _____ cm

  b. Line Z doubled would measure about _____ paper clips or about _____ cm long.

2. ——————————————————————————— Line A

                    ————————————————— Line B

  a. Line A

      _____ paper clips   _____ cm

  b. Line B

      _____ paper clips   _____ cm

  c. Line A is about _____ paper clips longer than Line B.

  d. Line B doubled is about _____ cm shorter than Line A.

Lesson 7: Measure and compare lengths using standard metric length units and non-standard length units; relate measurement to unit size.

31

EUREKA MATH

©2015 Great Minds. eureka-math.org
G2-M2-SE-B1-1.3.1-1.2016

3. Draw a line that is 9 cm long and another line below it that is 12 cm long.

   Label the 9 cm line F and the 12 cm line G.

   a. Line F   Line G

       _____ paper clips   _____ paper clips

   b. Line G is about _____ cm longer than Line F.

   c. Line F is about _____ paper clips shorter than Line G.

   d. Lines F and G are about _____ paper clips long.

   e. Lines F and G are about _____ centimeters long

4. Jordan measured the length of a line with large paper clips. His friend measured the length of the same line with small paper clips.

   a. About how many paper clips did Jordan use? _____ large paper clips

   b. About how many small paper clips did his friend use? _____ small paper clips

   c. Why did Jordan's friend need more paper clips to measure the same line as Jordan?

       _____

       _____

Lesson 7:   Measure and compare lengths using standard metric length units and non-standard length units; relate measurement to unit size.

EUREKA MATH

Name _____   Date _____

1.

                 A   _____

      B   _____

```
|   |   |   |   |   |   |   |   |   |   |   |   |   |   |   |
  1   2   3   4   5   6   7   8   9   10  11  12  13  14  15
```

  a. Line A is _____ cm long.

  b. Line B is _____ cm long.

  c. Together, Lines A and B measure _____ cm.

  d. Line A is _____ cm (longer/shorter) than Line B.

2. A cricket jumped 5 centimeters forward and 9 centimeters back, and then stopped.
    If the cricket started at 23 on the ruler, where did the cricket stop? Show your
    work on the broken centimeter ruler.

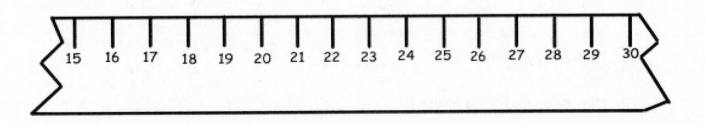

3. Each of the parts of the path below is 4 length units. What is the total length of the path?

   _____ length units

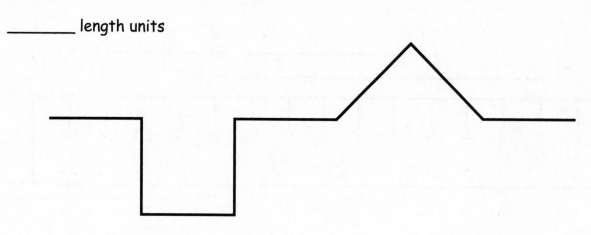

4. Ben took two different ways home from school to see which way was the quickest. All streets on Route A are the same length. All streets on Route B are the same length.

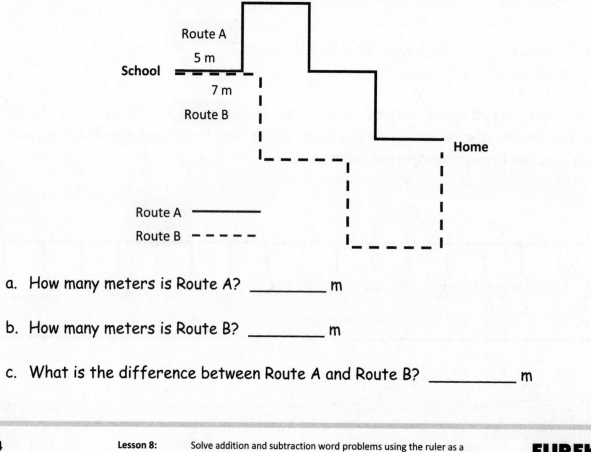

   a. How many meters is Route A? _____ m

   b. How many meters is Route B? _____ m

   c. What is the difference between Route A and Route B? _____ m

Lesson 8:   Solve addition and subtraction word problems using the ruler as a number line.

EUREKA MATH

Name _____   Date _____

1.

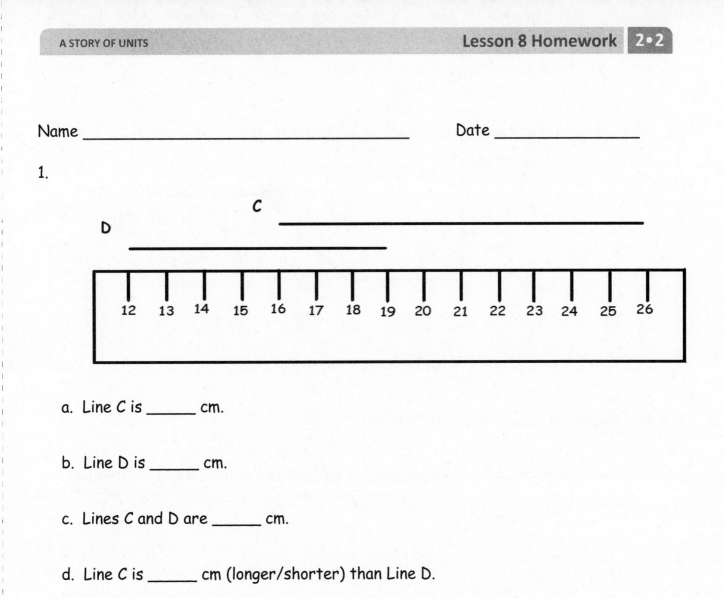

a. Line C is _____ cm.

b. Line D is _____ cm.

c. Lines C and D are _____ cm.

d. Line C is _____ cm (longer/shorter) than Line D.

2. An ant walked 12 centimeters to the right on the ruler and then turned around and walked 5 centimeters to the left. His starting point is marked on the ruler. Where is the ant now? Show your work on the broken ruler.

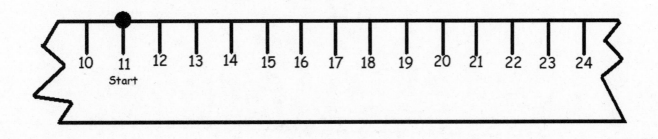

       Lesson 8:   Solve addition and subtraction word problems using the ruler as a   **35**
                               number line.

©2015 Great Minds. eureka-math.org
G2-M2-SE-B1-1.3.1-1.2016

3. All of the parts of the path below are equal length units.

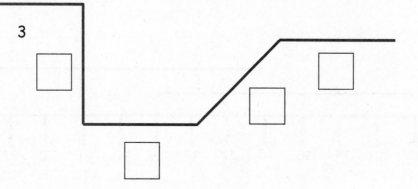

a. Fill in the empty boxes with the lengths of each side.

b. The path is _____ length units long.

c. How many more parts would you need to add for the path to be 21 length units long?

_____ parts

4. The length of a picture is 67 centimeters. The width of the picture is 40 centimeters. How many more centimeters is the length than the width?

Lesson 8:   Solve addition and subtraction word problems using the ruler as a number line.

Name _____   Date _____

1. Complete the chart by first estimating the measurement around a classmate's body
    part and then finding the actual measurement with a meter strip.

Student Name	Body Part Measured	Estimated Measurement in Centimeters	Actual Measurement in Centimeters
	Neck		
	Wrist		
	Head		

a. Which was longer, your estimate or the actual measurement around your

    classmate's head? _____

b. Draw a tape diagram to compare the lengths of two different body parts.

Lesson 9:   Measure lengths of string using measurement tools, and use tape
diagrams to represent and compare lengths.   37

©2015 Great Minds. eureka-math.org
G2-M2-SE-B1-1.3.1-1.2016

2. Use a string to measure all three paths.

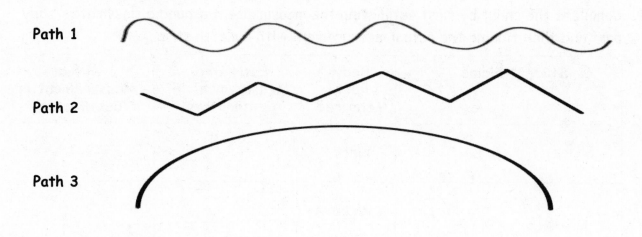

Path 1

Path 2

Path 3

a. Which path is the longest? _____

b. Which path in the shortest? _____

c. Draw a tape diagram to compare two of the lengths.

Measure lengths of string using measurement tools, and use tape
diagrams to represent and compare lengths.

©2015 Great Minds. eureka-math.org
G2-M2-SE-B1-1.3.1-1.2016

3. Estimate the length of the path below in centimeters.

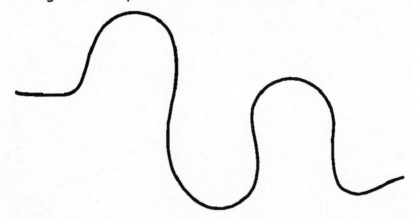

a. The path is about _____ cm long.

Use your piece of string to measure the length of the path. Then, measure the string with your meter strip.

b. The actual length of the path is _____ cm.

c. Draw a tape diagram to compare your estimate and the actual length of the path.

This page intentionally left blank

Name _____ Date _____

Use the RDW process to solve. Draw a tape diagram for each step. Problem 1 has been started for you.

1. Maura's ribbon is 26 cm long. Colleen's ribbon is 14 cm shorter than Maura's ribbon. What is the total length of both ribbons?

   Step 1: Find the length of Colleen's ribbon.

   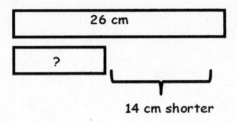

   Step 2: Find the length of both ribbons.

   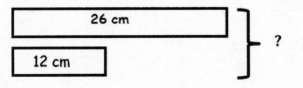

Lesson 10: Apply conceptual understanding of measurement by solving two-step word problems.

©2015 Great Minds. eureka-math.org
G2-M2-SE-B1-1.3.1-1.2016

43

2. Jesse's tower of blocks is 30 cm tall. Sarah's tower is 9 cm shorter than Jessie's tower. What is the total height of both towers?

   Step 1: Find the height of Sarah's tower.

   Step 2: Find the height of both towers.

3. Pam and Mark measured the distance around each other's wrists. Pam's wrist measured 10 cm. Mark's wrist measured 3 cm more than Pam's. What is the total length around all four of their wrists?

   Step 1: Find the distance around both Mark's wrists.

   Step 2: Find the total measurement of all four wrists.

Lesson 10:   Apply conceptual understanding of measurement by solving two-step
                            word problems.

Eureka Math
Grade 2
Module 3

Special thanks go to the Gordon A. Cain Center and to the Department of Mathematics at Louisiana State University for their support in the development of *Eureka Math*.

For a free *Eureka Math* Teacher
Resource Pack, Parent Tip
Sheets, and more please
visit www.Eureka.tools

Printed in the U.S.A.
This book may be purchased from the publisher at eureka-math.org
10 9 8

ISBN 978-1-63255-293-8

Name _____   Date _____

 Draw models of ones, tens, and hundreds. Your teacher will tell you which numbers to model.

Lesson 1:   Bundle and count ones, tens, and hundreds to 1,000.   **1**

©2015 Great Minds. eureka-math.org
G2-M3-SE-B1-1.3.1-1.2016

This page intentionally left blank

Name _____   Date _____

1. Draw, label, and box 100. Draw pictures of the units you use to count from 100 to 124.

_____

2. Draw, label, and box 124. Draw pictures of the units you use to count from 124 to 220.

3. Draw, label, and box 85. Draw pictures of the units you use to count from 85 to 120.

---

4. Draw, label, and box 120. Draw pictures of the units you use to count from 120 to 193.

Lesson 2:   Count up and down between 100 and 220 using ones and tens.

©2015 Great Minds. eureka-math.org
G2-M3-SE-B1-1.3.1-1.2016

Name _____   Date _____

1. Draw, label, and box 90. Draw pictures of the units you use to count from 90 to 300.

2. Draw, label, and box 300. Draw pictures of the units you use to count from 300 to 428.

EUREKA MATH

Lesson 3:   Count up and down between 90 and 1,000 using ones, tens, and hundreds.

©2015 Great Minds. eureka-math.org
G2-M3-SE-B1-1.3.1-1.2016

9

3. Draw, label, and box 428. Draw pictures of the units you use to count from 428 to 600.

4. Draw, label, and box 600. Draw pictures of the units you use to count from 600 to 1,000.

      Lesson 3:   Count up and down between 90 and 1,000 using ones, tens, and hundreds.

Name _____   Date _____

1. Fill in the blanks to reach the benchmark numbers.

   a. 14, _____, _____, _____, _____, _____, 20, _____, _____, 50

   b. 73, _____, _____, _____, _____, _____, _____, 80, _____, 100, _____, 300, _____, 320

   c. 65, _____, _____, _____, 70, _____, _____, 100

   d. 30, _____, _____, _____, _____, _____, 100, _____, _____, 400

2. These are ones, tens, and hundreds. How many sticks are there in all?

   There are _____ sticks in all.

3. Show a way to count from 668 to 900 using ones, tens, and hundreds.

Lesson 3:   Count up and down between 90 and 1,000 using ones, tens, and
             hundreds.

11

©2015 Great Minds. eureka-math.org
G2-M3-SE-B1-1.3.1-1.2016

4. Sally bundled her sticks in hundreds, tens, and ones.

a. How many sticks does Sally have? _____

b. Draw 3 more hundreds and 3 more tens. Count and write how many sticks Sally has now.

Lesson 3:   Count up and down between 90 and 1,000 using ones, tens, and
                                    hundreds.

 EUREKA MATH™

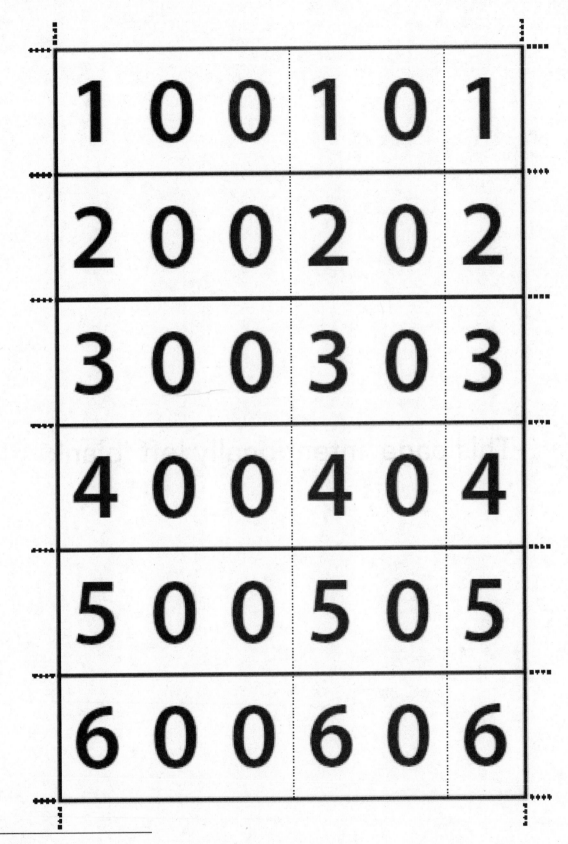

hide zero cards

©2015 Great Minds. eureka-math.org
G2-M3-SE-B1-1.3.1-1.2016

This page intentionally left blank

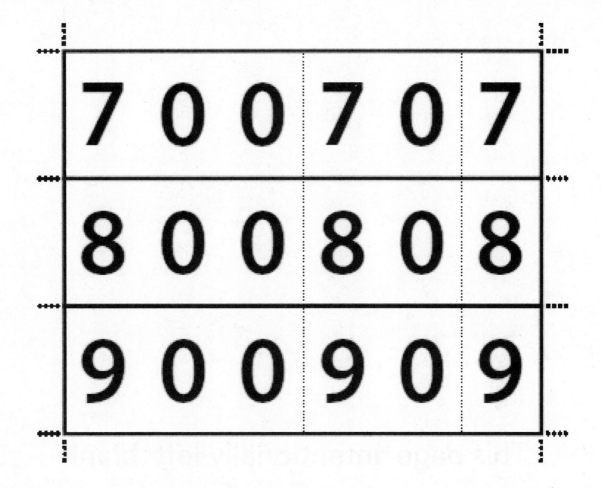

7 0 0 7 0 7

8 0 0 8 0 8

9 0 0 9 0 9

_____

hide zero cards

This page intentionally left blank

Name _____   Date _____

Your teacher will tell you a number to write in each box. In a whisper voice, say each number in word form. Use number bonds to show how many ones, tens, and hundreds are in the number.

Lesson 5:   Write base ten three-digit numbers in unit form; show the value of
            each digit.

23

©2015 Great Minds. eureka-math.org
G2-M3-SE-B1-1.3.1-1.2016

This page intentionally left blank

Name _____ Date _____

1. What is the value of the 7 in | 7 | 6 | 4 | ? _____

2. Make number bonds to show the hundreds, tens, and ones in each number. Then, write the number in unit form.

    a. 333

    ┌───┐
    │ Example: (263)   │
    │   ╱ │ ╲   │
    │   200 60   3   │
    │   │
    │   2 hundreds 6 tens 3 ones   │
    │   _____   │
    └───┘

    _____

    b. 330

    _____

    c. 303

    _____

EUREKA MATH™

Lesson 5:   Write base ten three-digit numbers in unit form; show the value of each digit.

25

3. Draw a line to match unit form with number form.

a. 1 hundred 1 one =   11

b. 1 ten 1 one =   710

c. 7 tens 1 one =   110

d. 7 hundreds 1 one =   701

e. 1 hundred 1 ten =   101

f. 7 hundreds 1 ten =   71

Lesson 5:   Write base ten three-digit numbers in unit form; show the value of each digit.

ones	
tens	
hundreds	

ones	
tens	
hundreds	

ones	
tens	
hundreds	

ones	
tens	
hundreds	

individual place value charts

This page intentionally left blank

Name _____   Date _____

Spell Numbers: How many can you write correctly in 2 minutes?

1		11		10	
2		12		20	
3		13		30	
4		14		40	
5		15		50	
6		16		60	
7		17		70	
8		18		80	
9		19		90	
10		20		100	

_____

mber spelling activity sheet

Lesson 7:   Write, read, and relate base ten numbers in all forms.

©2015 Great Minds. eureka-math.org
G2-M3-SE-B1-1.3.1-1.2016

This page intentionally left blank

Name _____   Date _____

Show each amount of money using 10 bills: $100, $10, and $1 bills. Whisper and write each amount of money in expanded form. Write the total value of each set of bills as a number bond.

10 Bills

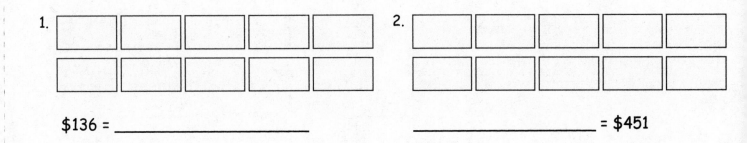

1.

$136 = _____

2.

_____ = $451

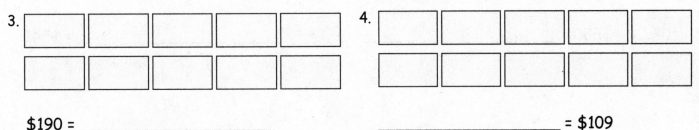

3.

$190 = _____

4.

_____ = $109

EUREKA
MATH™

Lesson 8:   Count the total value of $1, $10, and $100 bills up to $1,000.

5.

$460 = _____

6.

_____ = $406

7.

$550 = _____

8.

_____ = $541

EUREKA
MATH™

9.

$901 = _____

10.

_____ = $910

11.

$1,000 = _____

12.

_____ = $100

Lesson 8:   Count the total value of $1, $10, and $100 bills up to $1,000.

41

©2015 Great Minds. eureka-math.org
G2-M3-SE-B1-1.3.1-1.2016

This page intentionally left blank

Name _____   Date _____

1. Write the total value of the money.

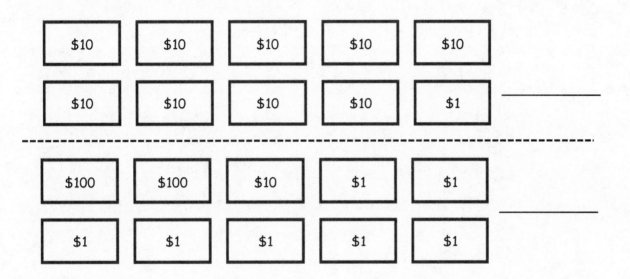

2. Fill in the bills with $100, $10, or $1 to show the amount.

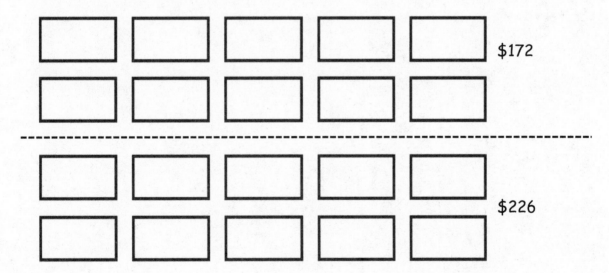

3. Draw and solve.

Brandon has 7 ten-dollar bills and 8 one-dollar bills. Joshua has 3 fewer ten-dollar bills and 4 fewer one-dollar bills than Brandon. What is the value of Joshua's money?

Lesson 8:   Count the total value of $1, $10, and $100 bills up to $1,000.

unlabeled hundreds place value chart

Lesson 8:   Count the total value of $1, $10, and $100 bills up to $1,000.

45

©2015 Great Minds. eureka-math.org
G2-M3-SE-B1-1.3.1-1.2016

This page intentionally left blank